I0066998

Henry James

Plans and photographs of Stonehenge and of Turusachan in the Island of Lewis

With notes relating to the Druids and sketches of cromlechs in Ireland

Henry James

Plans and photographs of Stonehenge and of Turusachan in the Island of Lewis
With notes relating to the Druids and sketches of cromlechs in Ireland

ISBN/EAN: 9783741122859

Manufactured in Europe, USA, Canada, Australia, Japa

Cover: Foto ©berggeist007 / pixelio.de

Manufactured and distributed by brebook publishing software
(www.brebook.com)

Henry James

Plans and photographs of Stonehenge and of Turusachan in the Island of Lewis

PLANS AND PHOTOGRAPHS

OF

STONEHENGE,

AND OF

TURUSACHAN IN THE ISLAND OF LEWIS;

WITH

NOTES RELATING TO THE DRUIDS

AND

SKETCHES OF CROMLECHS IN IRELAND,

BY

COLONEL SIR HENRY JAMES, R.E., F.R.S., F.G.S., M.R.I.A., &c.

DIRECTOR-GENERAL OF THE ORDNANCE SURVEY.

1867.

PREFACE.

This short account of Stonehenge and Turusachan, with the few well-known passages from ancient authors relating to the Druids, and to the progress made in the mechanical arts in Gaul and Britain, at, and for some time before the Roman conquest, is circulated for the information of the Officers on the Ordnance Survey, in the hope that it may stimulate them to make Plans and Sketches, and to give Descriptive Remarks of such Objects of Antiquity as they may meet with during the progress of the Survey of the Kingdom. I have also given Sketches of four Irish Cromlechs, for comparison with those found in Great Britain.

HENRY JAMES,

Colonel, Royal Engineers,
Director-General of the Ordnance Survey.

SOUTHAMPTON,
29th May, 1867.

STONEHENGE.

The celebrated structure called Stonehenge, that is, the "Hanging Stones," stands in a commanding position on Salisbury Plain, about seven and a-half miles north of the city of Salisbury and two miles west of Amesbury.

The plain for two or three miles round Stonehenge is thickly studded over with tumuli, in almost every one of which cinerary urns containing the calcined bones of the dead, with their ornaments and arms, have been found; and the whole plain to the distance of from ten to fifteen miles round, is more or less covered with similar tumuli.

The structure, when complete, consisted of an outer circle of thirty large stones, upon which thirty other large stones were laid horizontally so as to form a perfect continuous circle. This circle is 100 feet in diameter within the stones.

The stones in the uprights have each two tenons on their upper surfaces, which fit into mortices cut into the under surface of the horizontal stones, by this mode of construction the whole circle was braced together. The average dimensions of the uprights in this circle are 12 ft. 7 in. high out of the ground, 6 ft. broad, and 3 ft. 6 in. thick.

B

Those in the circle resting on the uprights are about 10 ft. long, 3 ft. 6 in. wide, and 2 ft. 8 in. deep.

Within this circle there are five stupendously large trilithons, each consisting of two uprights with tenons on them, supporting a large horizontal lintel, in which two mortices are cut to receive the tenons.

These trilithons, as may be seen on the plan, are arranged in the form of a horse shoe, so that one of them at A is central as regards the other four. The horizontal stone F called the Altar stone, lies in front of the central trilithon, and we see that the axial line of the structure is from N.E. to S.W., or on the line of the two stones G, H. The five trilithons are arranged very symmetrically within the outer circle, and nearly at the distance of half the radius from the centre.

The dimensions of the trilithons are nearly as follows:—

	feet	in.		feet	in.		feet	in.
A—Height of upright out of ground,	22	5	Breadth,	7	6	Thickness,	4	0
Lintel, Length,	15	0	,,	4	6	,,	3	6
B—Height of upright out of ground,	17	2	,,	7	0	,,	4	0
Lintel, Length,	15	9	,,	4	0	,,	3	7
C—Height of upright out of ground,	16	6	,,	7	9	,,	4	0
Lintel, Length,	17	0	,,	4	0	,,	2	8
D—Height of upright out of ground,	22	0	,,	8	3	,,	4	3
Lintel, Length,	16	0	,,	4	0	,,	3	6
E—Height of upright out of ground,	16	6	,,	7	0	,,	4	0

The Altar stone F is 17 ft. long and 3 ft. 6 in. wide.

All the stones in the outer circle and in the trilithons are of an indurated tertiary sandstone, which is found upon the chalk in the neighbourhood, and more particularly near Avebury and Marlborough, where they are known by the name of the " Sarsen " stones, and the " Grey Wethers."

They have all been roughly squared and dressed, especially at the joints between the lintels and uprights where the surfaces have been truly worked and the tenons and mortices truly fitted into each other.

In addition to these there was formerly a complete circle of thirty smaller upright stones about six feet high, which was intermediate in position between the outer circle and the five trilithons. Within the trilithons there was also a row of smaller stones about seven feet six inches high, parallel to the trilithons as represented on the plan of " Stonehenge restored."

Mr. Cunnington in a letter to Sir Richard Colt Hoare, has suggested the idea that these small stones did not form part of the original structure, an opinion in which I concur. They were probably monumental stones to the memory of chieftains and priests, afterwards erected within the temple.

These smaller stones are of various kinds of igneous and primitive rocks, and are evidently erratic blocks from the north of England and from Scotland, transported by the agency of ice. Many of such stones are still to be seen lying about in all parts of the country, although for ages the people have been using them for building and other purposes.

Only seventeen of the thirty upright stones of the outer circle are now standing, and only six of the thirty lintels are now in their places. Of the trilithons only two (B and C) are perfect ; the lintel and one of the uprights of A has fallen and lies broken upon the Altar stone F, whilst the other upright is in an inclined position, and supported only by one of the smaller stones which stood in front of it; this fell in 1620. D lies prostrate, having fallen outward with its capstone on the 3rd January, 1797. One of the uprights of E has fallen inwards and is broken into three parts, and its lintel also is broken into three parts.

Of the circle of smaller stones very few remain standing; the small lintel on the left of the central entrance is all that remains to indicate that there were probably some lintels on this circle, as there may also have been on the inner row of stones.

The structure is surrounded by a circular enclosure of earth, about 300 feet in diameter, with a shallow ditch outside it.

The construction of this grand work has traditionally, and I think rightly, been attributed to the Druids, of whom a short account is given below.

Those who require a fuller account of Stonehenge should consult Sir Richard Colt Hoare's great work on the Antiquities of Wiltshire.

THE DRUIDS.

DIODORUS SICULUS, who wrote about forty years before Christ, tells us, although he distrusts the accounts, regarding them "as much like fables," that Hecatæus, [who wrote about 500 years before Christ] and others, say, "there is an "island in the ocean over against Gaul, (as big as Sicily) under the arctic pole, "that the soil here is very rich, and very fruitful, and the climate temperate, "insomuch as there are two crops in the year," and that "the Hyperboreans "inhabit this island, and use their own natural language."

It is curious that from his own description of Britain, which he states is "over against Gaul," "in form triangular like Sicily," as being "under the arctic pole," and "the inhabitants the original people thereof," Diodorus did not recognise the fact that this island of the Hyperboreans was in reality the island of Britain. Indeed there is no other island to which this description could by possibility apply.

Hecatæus wrote at a time when the Western parts of Europe were almost absolutely unknown to the Greeks, and only such imperfect knowledge reached them as was brought by adventurous traders in gold, tin, amber, &c., and these we know from history were very jealous to conceal the knowledge of the places from whence they brought these valuable products. Thus Herodotus, who wrote about the year 410 B.C., or 90 years after Hecatæus, whose writings he quotes, says :—

"Concerning the Western extremities of Europe I am unable to speak with certainty, for I do not admit that there is a river called by the barbarians Eridanus (the Rhine) which discharges itself into the sea towards the north, from which amber is said to come, nor am I acquainted with the Cassiterides islands from whence our tin comes. For in the first place, the name Eridanus shows that it is Grecian and not barbarian, and feigned by some poet; in the next place, though I have diligently enquired, I have never been able to hear from any man who has himself seen it, that there is a sea on that side of Europe. However, both tin and amber come to us from the remotest parts. Towards the north of Europe there is evidently a great quantity of gold, but how procured I am unable to say with certainty."—iii. 115.

We now know that the accounts which Herodotus and Diodorus so much distrusted were correct; but seeing how very limited was the knowledge of the Geography of Europe which they possessed, we need not be surprised at their incredulity. This island was called the island of the Hyperboreans under the idea that it was in "the void recesses of Nature," and beyond the place from which the north winds came, just as the name of Hypernotians was given to those who lived beyond where the south winds were supposed to turn back again.

Diodorus, quoting from Hecatæus goes on to say "these inhabitants demean themselves as if they were Apollo's priests, who has there a STATELY GROVE and RENOWNED TEMPLE OF A ROUND FORM, beautified with many rich gifts;"—and I incline to the belief, held by many writers, that the island here referred to could be no other than Britain, and that this celebrated temple of a round form is no other than Stonehenge, and that the priests referred to were the Druids.

That the island of the Hyperboreans was Britain, may be also argued from Cæsar's account of the Druids in Gaul, in which he says that "their institution "came originally from Britain, and even at this day such as are desirous of being "perfect in it travel thither for instruction." Britain, therefore, was clearly considered to be the school and head quarters of the Druidical order, and would have been so quoted by writers such as Hecatæus, under whatever fanciful name their ignorance of geography might lead them to adopt. For information more or less exact respecting the Druids of Britain and their temples would through the travellers mentioned by Cæsar, be spread over the whole of Gaul, from the Atlantic to the Adriatic, and so carried on to Greece. If, therefore, we grant that the island of the Hyperboreans was Britain, it is almost impossible to resist the inference that the "renowned temple" of the Druids mentioned by Hecatæus was Stonehenge, for neither in Britain or elsewhere in Europe is any structure to be found equal to it in magnitude and design.

The vast number of tumuli which surround Stonehenge, containing in rude urns the ashes of deceased chieftains and men of note, with their arms and personal ornaments, prove that this must have been a celebrated place of worship and burial long before the Roman conquest of Britain. These tumuli and their contents, as Sir R. C. Hoare observes, "evidently prove their high antiquity, and mark them of an era prior to the Roman invasion."

From the intimate relations existing between the two countries, and their inhabitants being of one religion, we may safely assume that the description given of the Druids in Gaul also applies to those who at the same time resided in Britain.

The intimate relations which existed between Gaul and Britain at the time of, and before the invasion of Cæsar is obvious from what he says of the Suessiones who were "possessed of a very large and fruitful country, over which, even of late "years, Divitiacus had been king, one of the most powerful princes of all Gaul, "and who, besides his dominions in those parts, reigned also over Britain."— Book III. Chap. 4.

In speaking of the Veneti, whose territories were on the coast about Cherbourg, against whom he was preparing an expedition, he says:—

"This last state is by far the most powerful and considerable of all the nations inhabiting along the sea coast; and that not only on account of their vast shipping, wherewith they drive a mighty traffic in Britain, and their skill and experience in naval affairs, in which they greatly surpass the other maritime states; but because lying upon a large and open coast, against which the sea rages with great violence, and where the havens, being few in number, are all subject to their jurisdiction; they have most of the nations that trade in those seas tributary to their state."—Book III., Chap. 8.

The Veneti, he further says: Chap. 9, "despatched ambassadors into Britain, "which lies over against their coast, to solicit assistance from thence." Again, in Book IV. Chap. 18 he says that he "resolved to pass over into Britain, having "certain intelligence that in all his wars with the Gauls, the enemies of the "commonwealth had ever received assistance from thence;" and in Book V. Chap. 10, describing Britain, he says: "The island is well peopled, full of houses "built after the manner of the Gauls, and abounds in cattle. They use brass "money and iron rings of a certain weight. The provinces remote from the sea "produce tin, and those upon the coast iron."

It is worthy of remark that the Veneti also possessed the country about Quiberon on the Bay of Biscay, which includes the extraordinary Druidical structure at Carnac, in which there are 4000 huge upright unwrought stones.

The following is the account of the Druids given by Cæsar [about 50 B.C.], from which the accounts of subsequent writers are principally taken :—

"Over all Gaul, there are only two orders of men, in any degree of honour and esteem: for the common people are little better than slaves, attempt nothing of themselves, and have no share in the public deliberations. As they are generally oppressed with debt, heavy tributes, or the exactions of their superiors, they make themselves vassals to the great, who exercise over them the same jurisdiction as masters do over slaves. The two orders of men, with whom, as we have said, all authority and distinctions are lodged, are the Druids and nobles. The Druids preside in matters of religion, have the care of public and private sacrifices, and interpret the will of the gods. They have the direction and education of the youth, by whom they are held in great honour. In almost all controversies, whether public or private, the decision is left to them: and if any crime is committed, any murder perpetrated; if any dispute arises touching an inheritance, or the limits of adjoining estates; in all such cases, they are the supreme judges. They decree rewards and punishments; and if any one refuses to submit to their sentence, whether magistrate or private man, they interdict him the sacrifices. This is the greatest punishment that can be inflicted among the Gauls; because such as are under this prohibition, are considered as impious and wicked: all men shun them, and decline their conversation and fellowship, lest they should suffer from the contagion of their misfortunes. They can neither have recourse to the law for justice, nor are capable of any public office. The Druids are all under one chief, who possesses the supreme authority in that body. Upon his death, if any one remarkably excels the rest, he succeeds; but if there are several candidates of equal merit, the affair is determined by plurality of suffrages. Sometimes they even have recourse to arms before the election can be brought to an issue. Once a year they assemble at a consecrated place in the territories of the Carnutes,* whose country is supposed to be in the middle of Gaul. Hither such as have any suits depending, flock from all parts, and submit implicitly to their decrees. Their institution is supposed to come originally from Britain, whence it passed into Gaul; and even at this day, such as are desirous of being perfect in it, travel thither for instruction. The Druids never go to war, are exempted from taxes and military service, and enjoy all manner of immunities. These mighty encouragements induce multitudes of their own accord to follow that profession; and many are sent by their parents and relations. They are taught to repeat a great number of verses by heart, and often spend twenty years upon this institution; for it is deemed unlawful to commit their statutes to writing; though in other matters, whether public or private, they make use

* At Carnac?

of Greek characters. They seem to me to follow this method for two reasons: to hide their mysteries from the knowledge of the vulgar; and to exercise the memory of their scholars, which would be apt to lie neglected, had they letters to trust to, as we find is often the case. It is one of their principal maxims that the soul never dies, but after death passes from one body to another; which, they think, contributes greatly to exalt men's courage, by disarming death of its terrors. They teach likewise many things relating to the stars and their motions, the magnitude of the world and our earth, the nature of things, and the power and prerogatives of the immortal gods.

"The other order of men is the nobles, whose whole study and occupation is war. Before Cæsar's arrival in Gaul, they were almost every year at war either offensive or defensive; and they judge of the power and quality of their nobles, by the vassals, and the number of men he keeps in his pay; for they are the only marks of grandeur they make any account of.

"The whole nation of the Gauls is extremely addicted to superstition: whence, in threatening distempers, and the imminent dangers of war, they make no scruple to sacrifice men, or engage themselves by vow to such sacrifices; in which they make use of the ministry of the Druids: for it is a prevalent opinion among them, that nothing but the life of man can atone for the life of man; insomuch that they have established even public sacrifices of this kind. Some prepare huge Colossuses, of osier twigs, into which they put men alive, and setting fire to them, those within expire amidst the flames. They prefer for victims such as have been convicted of theft, robbery, or other crimes; believing them the most acceptable to the gods: but when real criminals are wanting, the innocent are often made to suffer. Mercury is the chief deity with them: of him they have many images, account him the inventor of all arts, their guide and conductor in their journeys, and the patron of merchandise and gain. Next to him are Apollo, and Mars, and Jupiter, and Minerva. Their notions in regard to him are pretty much the same with those of other nations. Apollo is their God of physic; Minerva of works and manufactures; Jove holds the empire of heaven; and Mars presides in war. To this last, when they resolve upon a battle, they commonly devote the spoil. If they prove victorious, they offer up all the cattle taken, and set apart the rest of the plunder in a place appointed for that purpose: and it is common in many provinces, to see these monuments of offerings piled up in consecrated places. Nay, it rarely happens that any one shows so great a disregard of religion, as either to conceal the plunder, or pillage the public oblations; and the severest punishments are inflicted upon such offenders.

"The Gauls fancy themselves to be descended from the god Pluto; which, it seems, is an established tradition among the Druids. For this reason they compute the time by nights, not by days; and in the observance of birth days, new moons, and the beginning of the year, always commence the celebration from the preceding night. In one custom they differ from almost all other nations; that they never suffer their children to come openly into their presence, until they are of age to bear arms: for the appearance of a son in public with his father, before he has reached the age of manhood, is accounted dishonourable.

"Whatever fortune the woman brings, the husband is obliged to equal it with his own estate. This whole sum, with its annual product, is left untouched, and falls always to the share of the survivor. The men have power of life and death over their wives and children; and when any father of a family of illustrious rank dies, his relations assemble, and upon the least ground of suspicion, put even his wives to the torture like slaves. If they are found guilty, iron and fire are employed to torment and destroy them. Their funerals are magnificent and sumptuous, according to their quality. Every thing that was dear to the deceased, even animals, are thrown into the pile: and formerly, such of their slaves and clients as they loved most, sacrificed themselves at the funeral of their lord."

DIODORUS SICULUS, Book v. Chap. 2 [B.C. 40] speaking of the Gauls, says :—

"There are likewise among them philosophers and divines whom they call Saronidæ (Druids), and who are held in great veneration and esteem.

c

"When they are to consult on some great and weighty matter, they observe a most strange and incredible custom, for they sacrifice a man, striking him with a sword near the diaphragm across over his breast, who being thus slain, and falling down, they judge of the event from the manner of his fall, the convulsion of his members, and the flux of blood; this has gained among them (by long and ancient usage) a firm credit and belief.

"It is not lawful to offer any sacrifice without a philosopher. These Druids and Bards are observed and obeyed, not only in time of peace but war also, both by friends and enemies.

"Malefactors they impale upon stakes, in honour to the Gods, and then with many other victims, upon a vast pile of wood, they offer them up as a burnt sacrifice to their deities..........In like manner they use their captives also, as sacrifices to the Gods."

STRABO, Book iv. Chap. iv. 4, 5 [B.C. 30]—

"Without the Druids they (the Gauls) never sacrifice........They would strike a man devoted as an offering in his back with a sword, and divine from his convulsive throes. It is said they have other modes of sacrificing their human victims; that they pierce some of them with arrows, and crucify others in their temples; and that they prepare a colossus of hay and wood, into which they put cattle, beasts of all kinds, and men, and then set fire to it."

PLINY, Book xvi. Chap. 95 [A.D. 75]—

"Upon this occasion we must not omit to mention the admiration that is lavished upon this plant (the mistletoe) by the Gauls. The Druids—for that is the name they give to their magicians—hold nothing more sacred than the mistletoe, and the tree that bears it, supposing always that tree to be the robur. Of itself the robur is selected by them to form whole groves, and they perform none of their religious rites without employing branches of it; so much so, that it is very probable that the priests themselves may have received their name from the Greek name (Drus, an oak) for that tree.

"The mistletoe, however, is but rarely found upon the robur; and when found, is gathered with rites replete with religious awe. This is done more particularly on the fifth day of the moon, the day which is the beginning of their months and years, as also of their ages, which with them are but thirty years. This day they select because the moon, though yet not in the middle of her course, has already considerable power and influence; and they call her by a name which signifies in their language, the all-healing. Having made all due preparation for the sacrifice and a banquet beneath the trees, they bring thither two white bulls, the horns of which are bound then for the first time. Clad in a white robe, the priest ascends the tree and cuts the mistletoe with a golden sickle, which is received by others in a white cloak. They then immolate the victims, offering up their prayers that God will render this gift propitious to those to whom he has so granted it."

PLINY, Book xxx. Chap. 4—

"The Gallic provinces, too, were pervaded by the magic art, and that even down to a period within memory; for it was the Emperor Tiberius that put down their Druids, and all that tribe of wizards and physicians. But why make further mention of these prohibitions, with reference to an art which has now crossed the very ocean over, and has penetrated to the void recesses of Nature? At the present day, struck with fascination, Britannia still cultivates this art, and that with ceremonials so august, that she might almost seem to have been the first to communicate them to the people of Persia."

SUETONIUS, in Claudius xxv. [A.D. 100]—

"The religious rites of the Druids, solemnized with such horrid cruelties, which had only been forbidden the citizens of Rome during the reign of Augustus, he utterly abolished among the Gauls."—p. 318.

Dion Chrysostomus [about A.D. 100] observes that "the Celtic kings could not so much as design any public measure without the Druids, who were adepts in divination and philosophy; insomuch that these priests exercised regal authority, and that the kings, who had but the semblance of power, were in truth their servants."

We thus see what enormous power the Druids possessed, and that not only were the inhabitants of the two countries taught in the same manner by them, but that as regards the state of the mechanical arts and manufactures of the day the two countries must have been nearly equally advanced—for when Cæsar tells that the Veneti had such an immense fleet that at one time "two hundred and "twenty of their best ships, well equipped for service and furnished with all "kinds of weapons, stood out to sea and drew up in order of battle against us." —Book III. Chap. 14, it may safely be inferred that the ships of their British allies were included in this number, and that they were built in the same manner. "The body of the vessels" he says "was entirely of oak, to stand the shocks "and assaults of that tempestuous ocean." The benches of the rowers were "made "of strong beams of about a foot in breadth, and fastened with iron nails an inch "thick, that instead of cables, they secured their anchors with chains of iron;" and as he also tells us their ships were large rather flat-bottomed sailing vessels, with very high bows and sterns, which gave them a great advantage over the Roman gallies, they must have been in fact in their outline very much like the modern Dutch galliot, and their chain cables must necessarily have been very strong, and thus we see that their knowledge of ship-building and of the mechanical arts, and especially the art of manufacturing iron in every required form must have been very great.

This more advanced state of the mechanical arts in Gaul and Britain than I think is generally admitted, enables us to understand how it was that the Britons were able to construct such an extraordinary number of chariots of war armed with scythes, that as Cæsar tells us, when Cassivelaunus the British king was disbanding his forces, he retained "only four thousand chariots." We should be justified in drawing the same inference as to their knowledge of the mechanical arts from the fact that for centuries before the invasion of Cæsar the inhabitants of this country were well versed in the art of mining for and smelting tin. The ingots of tin were carried at a very early period by the Phœnicians to Cadiz and to the East through the Straits of Gibraltar, and after the Roman conquest of Gaul and Britain, across Gaul to Marseilles. Through these channels, some knowledge of Britain, and of the institution of the Druids would be sure to reach the East.

With the knowledge we thus possess of the unlimited power of the Druids over the rulers and people of Britain, and of all the resources of the country in men, horses and all the requisite mechanical appliances for the construction of such a temple as that at Stonehenge, we can have little doubt but that the tradition respecting the Druids is correct, and that it is to them we are indebted for this "wonder of the west." It was not possible without such a combination of skill and power as they alone possessed that such huge stones could have been transported several miles, worked into their required forms, and raised into their places. —And even supposing that the assistance of the Veneti, with their chain cables and mechanical skill was required, the Druids could easily have commanded it, and sent some of their best artificers and mechanics with their chains and tools from Southampton or Poole to Stonehenge.

It must have required ages to enable the Veneti to so far advance the art of manufacturing iron as to produce large chain cables, and there is no anachronism, therefore, in attributing to the Gauls and Britons a high degree of mechanical skill long before the time of Cæsar or Hecatæus.

It is in vain we look to the Romans for an example of such a structure and it is in vain we search our subsequent history for anything like proof of the existence of such a combination of skill and power in any one of the rulers in this country, to whom we could attribute, with any shadow of probability, the construction of this great work. Nor was there any object, after the introduction of Christianity, for which such a structure could be required.[*]

The peculiarly open character of this imposing temple was well suited to the extraordinary religious rites of the Druids. The temple in fact was admirably adapted, both by its position and construction, to produce the greatest possible effect on a vast concourse of people, assembled to witness their dreadful sacrifices.

---

* " Soon after the fall of the great trilithon in 1797, Mr. Cunnington dug out some of the earth that had fallen into the excavation, and found a " fragment of fine black Roman pottery."—Sir R. C. Hoare, vol. I, page 150. And on other occasions mentioned by him fragments of Roman pottery, and of coarse half-baked pottery, have been found by excavating within and about the temple; but these prove nothing beyond the fact that the place was visited by those who used vessels made of these materials. We should find now in addition to these, pieces of glass, china, and tobacco pipes.

The Druids were as Cæsar tells us especially observant of the movements of the heavenly bodies; and the selection of the number of thirty for the stones of the outer circle at Stonehenge is supposed to have been made in consequence of its representing the number of days in their lunar month.

"All nations (says Newton) before the just length of the solar year was "known, reckoned months by the course of the moon, and years by the returns "of winter and summer, spring and autumn; and in making calendars for "their festivals, they reckoned thirty days to a lunar month, and twelve lunar "months to a year ,taking the nearest round numbers."—Chronology of Ancient Kingdoms.

It is quite possible that the number five was also selected for the number of the great trilithons within the circle, as being the number of days required to complete the year of twelve months of thirty days each; but this is mere conjecture.

Driven back before the Roman arms, the Druids sought refuge in the island of Anglesey, the conquest of which was made [A.D. 60] by Suetonius Paullinus, the then Roman governor of Britain. Tacitus, in his Annals Book xiv. Chap. 31, gives the following very graphic description of their appearance and conduct on the eve of the battle which decided their fate as regards every part of the country to which the Roman arms extended:—

"On the shore stood the forces of the enemy, a dense array of arms and men, with women darting through the ranks like furies, their dress funereal, their hair dishevelled, and carrying torches in their hands. The Druids around the host, pouring forth dire imprecations, with their hands uplifted towards the heavens, struck terror into the soldiers by the strangeness of the sight; insomuch that, as if their limbs were paralysed, they exposed their bodies to the weapons of the enemy, without an effort to move. Afterwards, at the earnest exhortations of the general, and from the effect of their own mutual importunities that they would not be scared by a rabble of women and fanatics, they bore down upon them, smote all that opposed them to the earth, and wrapped them in the flames themselves had kindled. A garrison was then established to overawe the vanquished, and the groves dedicated to sanguinary superstitions destroyed; for they deemed it acceptable to their deities to make their altars fume with the blood of captives, and to seek the will of the gods in the entrails of men."

But the Druids still held power amongst the Gaels of the highlands and Western islands of Scotland: "In the latter times of the Druidical order, all the principal families in the Hebrides had their Druidth, who foretold future events, and decided all causes, civil and ecclesiastical." They ultimately retired to Iona, where it is said their order was not quite extinct on the arrival of St. Columba into the Western islands in the sixth century, when, although their order was abolished, their temples, modified as we see at Callernish, in the Isle of Lewis, were used in the new form of worship, and also as halls of justice.

Of the personal appearance of the Druids of this country we have in the following account of the inhabitants of the Cassiterides, or tin islands, a description which is supposed to apply to them :—

They are inhabited by men "in black cloaks, clad in tunics reaching to the feet, girt about the breast, and walking with staves, thus resembling the Furies we see in tragic representations."—STRABO, Book III. Chap. v.

The Cassiterides have been erroneously supposed by the Emperor Napoleon, in his life of Julius Cæsar, and by others, to be the Scilly islands, but as there is not a particle of tin in those islands, or any traces of works in search of tin or any other metal in them, it is clear that the Scilly islands are not the tin islands, and as we know from Diodorus Siculus that the tin was then as now raised in Cornwall, and after being cast into the form of astragali, for the facility of transport both by water and land, was embarked from St. Michael's Mount (the Ictin or Tin port of Diodorus), this description of the inhabitants of the tin islands really applies to the inhabitants of Cornwall, and most probably to the Druids—see my "Note on the Block of Tin dredged up in Falmouth Harbour."

In Cornwall, the old workings, often with some of the miners' tools in them, are always spoken of as the "Jews workings;" and the very names of the town called Marazion, (that is, Looking to Zion,) and Market Jew Street, which is on the mainland close to St. Michael's Mount, is a strong confirmation of the fact that the people of the East traded to St. Michael's Mount for tin.

A great number of modern authors have written upon this subject, amongst whom are Camden, Audley, Stukely, Wood, Hoare, and Armstrong.

H. J.

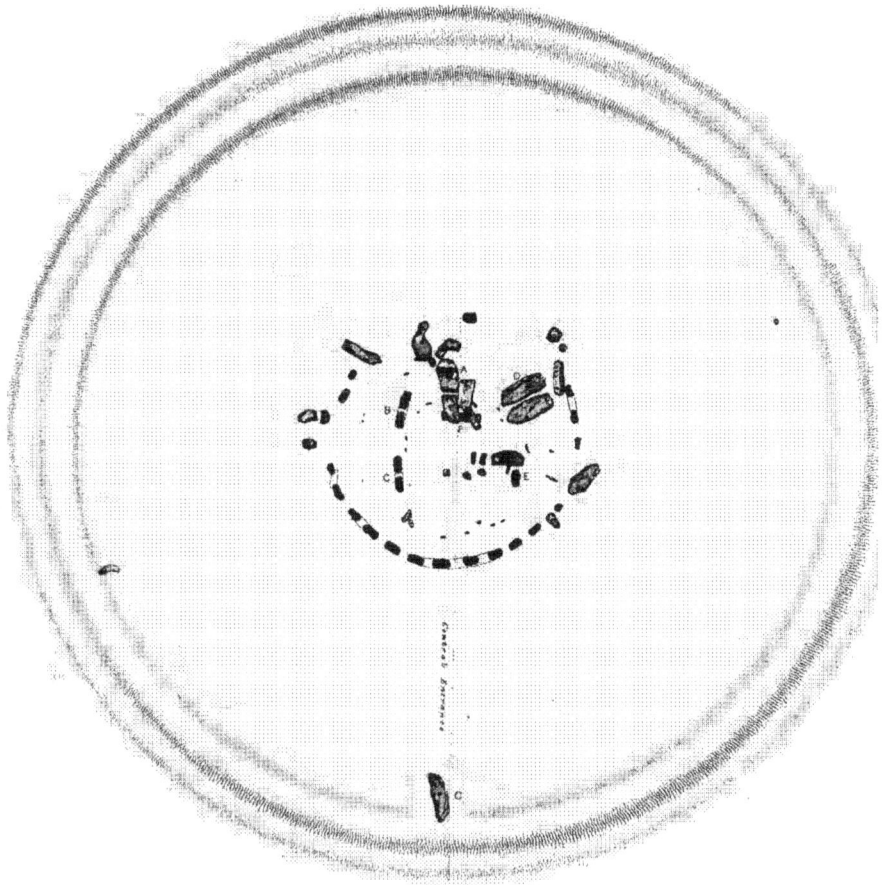

A B C D E TRILITHONS

F   ALTAR STONE

SCALE 1/600

50 40 30 20 10 0   50   100 FEET

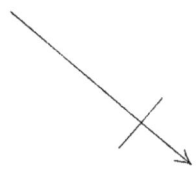H

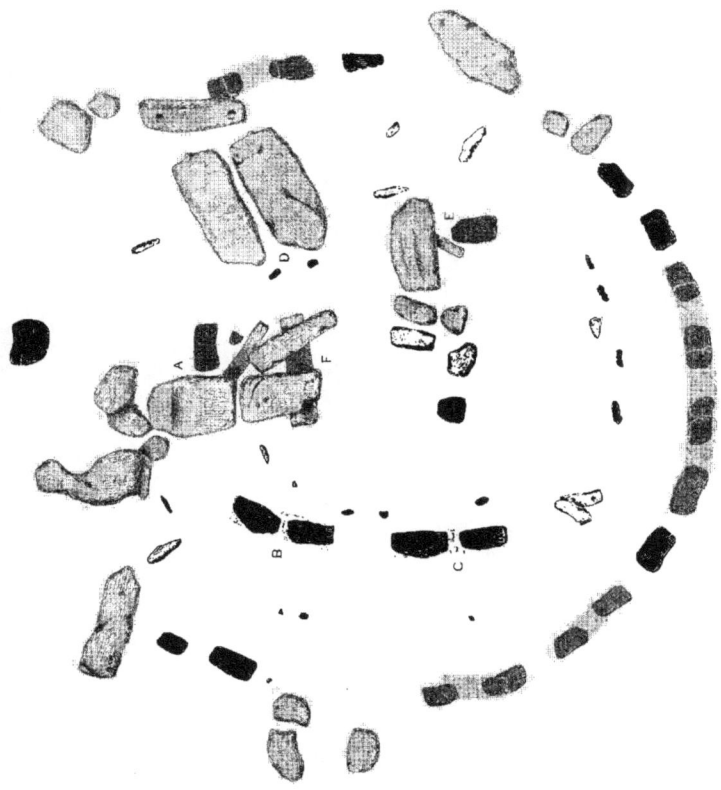

STONEHENGE

1867

Zincographed at the ORDNANCE SURVEY OFFICE, SOUTHAMPTON, Colonel Sir Henry James, R.E., F.R.S., Director.

SCALE

FEET

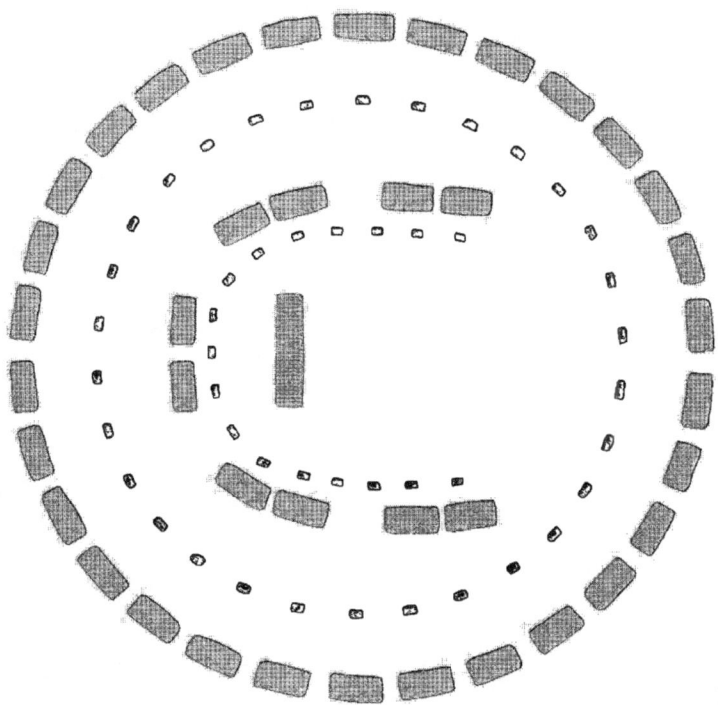

STONEHENGE
(restored)

SCALE

Feet

Zincographed at the ORDNANCE SURVEY OFFICE SOUTHAMPTON Colonel Sir Henry James R.E. F.R.S &c Director
1867.

STONEHENGE

CENTRE TRILITHON (A) AND ALTAR STONE UNDER THE LEFT UPRIGHT AND CAPSTONE OF THE TRILITHON

Photographed by the Ordnance Survey Department, Colonel Sir Henry James R.E. F.R.S &c Director.
1867.

STONEHENGE.

TRILITHON (B) ON THE LEFT OF ALTAR STONE.

Photographed by the Ordnance Survey Department. Colonel Sir Henry James R.E. F.R.S: &c Director
1867.

STONEHENGE.

TRILITHONS (B and C) from the SOUTH WEST.

Photographed by the Ordnance Survey Department, Colonel Sir Henry James R.E.: F.R.S &c Director.
1867

S T O N E H E N G E.

CENTRAL ENTRANCE through OUTER CIRCLE and UPRIGHT or TRILITHON (E).

Photographed by the Ordnance Survey Department Colonel Sir Henry James R.E. F.R.S. &c Director
1867.

STONEHENGE.

TRILITHONS E AND C AND PART OF OUTER CIRCLE

Photographed by the Ordnance Survey Department. Colonel Sir Henry James R.E. F.R.S.&c. Director
1867.

STONEHENGE

VIEW FROM SOUTH WEST

Photographed by the Ordnance Survey Department, Colonel Sir Henry James R.E. F R S. &c Director
1867.

STONEHENGE.

GENERAL VIEW FROM (G) ON THE LINE OF THE CENTRAL ENTRANCE

Photographed by the Ordnance Survey Department, Colonel Sir Henry James R.E. F.R.S. &c. Director.

1867

ACCOUNT OF THE DRUIDICAL STRUCTURE

CALLED

TURUSACHAN, AT CALLERNISH

IN THE

ISLAND OF LEWIS.

This very remarkable structure is thirteen miles due west of Stornaway, and is formed of forty-eight large upright stones, standing upon a low hill, the highest point of which is 143 feet above the mean level of the sea. This hill forms a peninsula at the bottom of East Loch Roag, and from its position with reference to this arm of the sea, and to West Loch Roag, it is much exposed to the westerly and north-westerly winds, and hence, probably, its descriptive name of Callernish, the Bleak or Cold Headland.

The structure itself is cruciform in plan, with a circle at the intersection of the shaft and arms, but with an additional row of stones on the east side of the shaft.

The circle, which is the great feature in the structure, is eighty feet above the level of the sea, and about forty-two feet in diameter, and consists of thirteen stones, with one in the centre which is seventeen feet high above the ground, and five feet six inches broad at the base.

From the centre stone to the northern extremity or foot of the shaft of the cross it is 294 feet, whilst the portion south of the circle is 114 feet, measured from the centre; the total length, therefore, is 408 feet. The extremity of the eastern arm is 73 feet, and of the western 57 feet from the centre; the total breadth of the cross is therefore 130 feet.

The stones are all of unwrought Gneiss, of which rock almost the whole island is formed.

This structure is described by Macculloch in his "Western Islands of Scotland," in which he says the centre stone is twelve feet high, as it must have been at the time when he wrote, but on the 2nd of October, 1857, Sir James Matheson, Bart., the enlightened and hospitable proprietor of the whole island, had the peat which had grown and accumulated over the hill and round the stones to the extent of five or six feet, entirely cleared away. The presence of the peat had prevented the growth of the moss and lichens on the stones, and the extent to which they had been buried can be plainly traced across them in the bleached appearance of their lower parts—as represented in the sketch.

The clearing away of the peat has brought to light the existence of what may be described as a cruciform grave, lying east and west, and so placed that the centre stone is made to serve as a head stone.

This grave is built of small stones, excepting at the four internal angles, which are formed by large single stones. This grave had been covered over with flat stones, but they had fallen in. It was built upon the natural surface of the ground, but the walls are supported by a bank of earth round them. Nothing was found in it when it was first opened.

It will be observed that the row of stones on the east side of those which form the shaft of the cross are so arranged as to form an avenue leading up to the grave.

The breadth of the centre portion of the grave is 5 ft. 9 in., and of the ends 2 ft.; the length of the centre 4 ft. 3 in., and of the west end 2 ft.; and the whole length about 10 ft., with a depth of about 2 ft.

There is an almost universal tradition amongst the Celtic races which connects the Druids, or priests of the ancient form of Pagan worship, with all the structures of this kind, and the preservation of such names as Clach an Druidean, the Stone of the Druids, in this island, would seem to prove their presence in it at some, possibly not very remote, period of our history.

As it is impossible that the turf could have been formed at the time these stones were set up, or that it could have grown during the time the place was much frequented, we may safely infer that many centuries must have elapsed since this structure was formed,—whilst the cruciform arrangement of the stones outside the circle, and the grave within it, and the avenue leading to the grave, seem to show that a great Pagan structure, having been adopted as a place of worship by Christians, had subsequently become venerated as the place of burial of a Saint, and so ultimately might become a place of pilgrimage, to be periodically visited by crowds of worshippers, as many places now are by Roman Catholics in all parts of the world.

The name Turus in the Irish language is always applied to the "stations" to which people resort on certain Saints' days to worship, and the presence of several other circles, also called Tursachan or Turnsachan, in the immediate neighbourhood of this great one seems to favour this idea, and hence "The place of pilgrimage" has been given as the translation of the Gaelic name Tnrusachan.

"The common Gaelic phrase—Am bheil thu dol dón chlachan—are you going to the stones?—by which the Scottish Highlander still enquires at a neighbour if he is bound for church, seems in itself no doubtful tradition of ancient worship within the monolithic ring."—Wilson's Archæology, page 110.

D

The name Callernish it is contended by some authorities should be Callanish, the place of assembly for worship, or calling to prayer, and that Turusachan should be Tursachan, as on the Ordnance Plan, i.e., the place of sadness, or sorrow, or of weariness.

These stones are known by the name of the Firbhreige, or false men, from their singular appearance at a distance on the crest of the hill, and their apparent movement amongst themselves as the observer changes his position.

HENRY JAMES,
Colonel, Royal Engineers,
Director-General of the Ordnance Survey.

Ordnance Survey Office,
29th May, 1867.

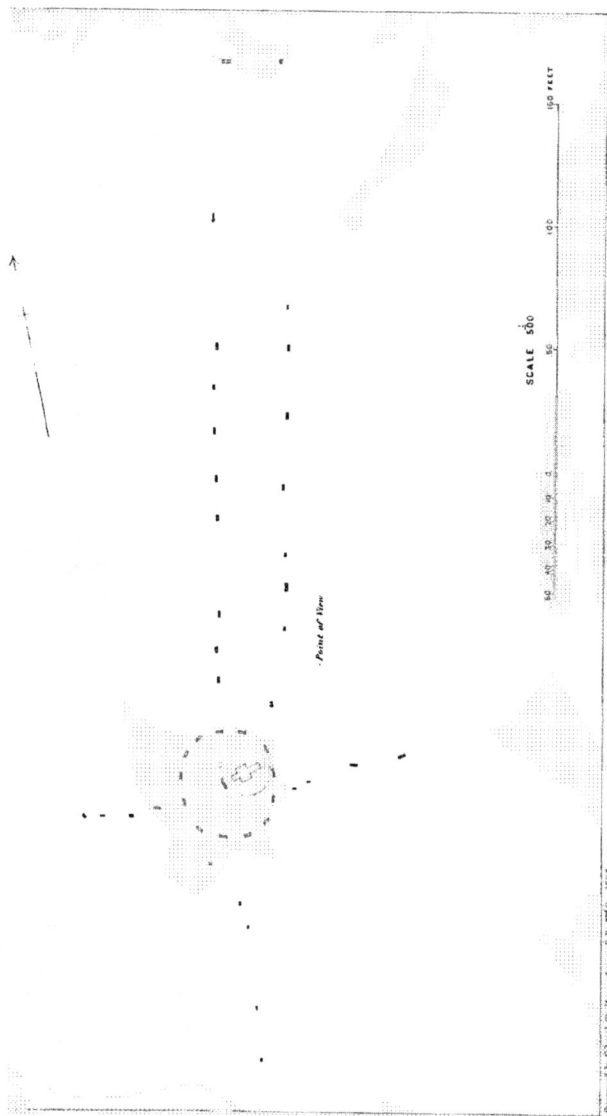

TURUSACHAN, CALLERNISH,

OR

THE PLACE OF PILGRIMAGE ON THE BLEAK HEADLAND.

IN THE

ISLE OF LEWIS.

Zincographed at the Ordnance Survey Office Southampton

1867.

SCALE 500

50   40   30   20   10   0   50   100   150 FEET

Point of View

Surveyed by Colonel Sir Henry James R.E. 3rd Sep 1858

TURUSACHAN. CALLERNISH.

OR

THE PLACE OF PILGRIMAGE ON THE BLEAK HEADLAND.

IN THE

ISLE OF LEWIS.

Drawn by Colonel Sir Henry James R.E. 1866

Copied in Chalk by 2nd Corporal Crawhan R.E.

SKETCHES OF

CROMLECHS IN IRELAND.

Since writing the short accounts of Stonehenge and Turusachan at Callernish, I have read in the "Athenæum" of the 20th of July, 1867, a review of Sir James Simpson's work on "Archaic Sculpturings," which contains the following extracts from that work:—

"In his interesting 'Himalayan Journal' (vol. II., p. 276), Dr. Hooker states that he found the Khasias, a wildish hill-tribe on the mountain confines of Upper India, still erecting megalithic structures. He remarks that among the Khasias, 'funeral ceremonies are the only ones of any importance, and they are often conducted with barbaric pomp and expense; and rude stones, of gigantic proportions, are erected as monuments, singly or in rows, or supporting one another, like those of Stonehenge, which they rival in dimensions and appearance.'"

"In reply to personal inquiries by Sir James Simpson, Dr. Hooker informed him—

"In answer to your query, Do you remember any *recent* erections, any arrangement the same as the cromlechs—viz., two, four, or six upright stones supporting a large mass?—this is the common erection now in vogue, such as are put up annually during the cold season. The whole country for many square miles was dotted with them, and they are annually put up. Some I saw were quite fresh, and others half-finished; and had I been there during the dry season, I was told I could have seen the operation. A chief or big man wants to put up such a cromlech, to commemorate an event or for any other purpose; he summons all the country-side, and feeds them for the time. They pass half the time in revelry, the other half in pulling, hauling, pushing, and prizing; it is all done by brute strength and stupidity. They have neither science nor craft, nor any implements of art but the lever. I was told that the ashes of the burnt dead were often deposited under them; but could not make out that this was a general custom. The whole country is studded with stone erections, usually a cromlech, with a row of tall stones behind it."

The first perusal of these extracts much surprised me, but it required very little reflection to see that although Stonehenge might be taken as a familiar object for comparison, and that although such a rude structure as that at Callernish might be erected by brute force, that such a symmetrical structure as that of Stonehenge, with its stones squared and dressed, and fitted together with tenons and mortices, was not erected by mere brute strength and stupidity, but by a people possessing great ability and mechanical skill. It will be interesting to have more particulars as to the manner in which the Khasias raise such heavy stones into their places, and I hope to obtain them through the Officers employed upon the Survey of India.

Some of the Cromlechs of Ireland may be also said with truth to rival Stonehenge in dimensions; and I have had the sketches of four of them, which I made many years ago, zincographed, to illustrate the character of these structures.

That in the Townland of Goard, between Castlewellan and Hilltown, in the County of Down, is remarkable for its size, the top stone being 14 ft. 3 in. long, with a mean width of 9 ft. 3 in., and 4 ft. 6 in. deep. It is of granite, and about 40 tons in weight. The lower end rests on a flat stone, which is supported by two upright stones, which form what may be presumed to be a grave. The highest point of the upper stone is 12 ft. 6 in. from the ground, and it is supported in front by two stones, one 7 ft. high, of porphyry, the other 5 ft. 10 in. high (See Plate 12). The jagged conical mountain, Slieve Bernia, in the Mourne Mountains, although visible all the way from Castlewellan to Hilltown, does not occupy the position given to it in the sketch, in fact, I have taken a painter's licence as regards the accessories of the sketch, but not with the Cromlech itself. This remark also applies to Nos. 14 and 15, but not to No. 13.

PLATE 13.—Represents the Cromlech in the Townland of Kilfeaghan, four miles and a half West of Kilkeel, in the County of Down. The upper stone is of granite, and about 30 tons in weight. It stands on the side of the hill overlooking Loch Carlingford.

PLATE 14.—Represents the Cromlech at Slidery Ford, near Dundrum Castle, in the County of Down. The upper stone is of granite. This is 7 feet from the top to the ground.

PLATE 15.—Represents the Cromlech at the Townland of Legannany, five miles North-west of Castlewellan, in the County of Down. The highest point is 8 ft. 6 in. from the ground; the upper corner of the lower end is 6 ft. from the ground. The upper stone is 11 ft. 3 in. long, 5 ft. 6 in. broad, and 2 ft. thick, but it is pointed at both ends, and may be said to be boat-shaped. It is supported on the points of three upright stones, and so nicely balanced that a boy can perceptibly move it.

These four may be said to represent the typical forms of the Irish Cromlechs, and of those found in Anglesey and other parts of the kingdom.

H. J.

CROMLECH

Six and a half miles South East of Castlewellan

CROMLECH

FOUR AND A HALF MILES WEST OF KILKEEL,

IN THE

COUNTY OF DOWN.

Sketched & Col. Sir H. James R.E. F.R.S. 8th June 1838

Drawn on Zinc by 2nd Corp! W. Crawforn R.E.

Zincographed at the ORDNANCE SURVEY OFFICE SOUTHAMPTON Colonel Sir Henry James R.E. F.R.S. &c Director

1867.

Sketched by Col. Sir H. James RE FRS 10th May 1828

Drawn on Zinc by 2nd Corp.l R.C.

CROMLECH

FIVE MILES NORTH WEST OF CASTLEWELLAN.

IN THE

COUNTY OF DOWN.

Sketched by col. Sir H. James R.E. F.R.S. 10th May 1853.

Drawn on Zinc by 2nd Corp.l W. Gairdner R.E.

Zincographed at the ORDNANCE SURVEY OFFICE SOUTHAMPTON Colonel Sir Henry James R.E. F.R.S. &c Director

1867

www.ingramcontent.com/pod-product-compliance
Lightning Source LLC
Chambersburg PA
CBHW022013190326
41519CB00010B/1509